Operation

Coding in the City

Scratch 3.0

By Kristin Fontichiaro and Colleen van Lent

Cherry Lake Press

Published in the United States of America by Cherry Lake Publishing
Ann Arbor, Michigan
www.cherrylakepublishing.com

Series Adviser: Kristin Fontichiaro
Reading Adviser: Marla Conn, MS, Ed., Literacy specialist, Read-Ability, Inc.

Image Credits: ©MissCreativa/Pixabay, 4, 6, 20; ©Clker-Free-Vector-Images/Pixabay, 4, 6, 8, 10, 12, 14, 16, 20; ©lethutrang101/Pixabay, 6, 8, 10, 12, 14, 16; ©Kaz/Pixabay, 6, 8, 10, 12, 14, 16; ©WildOne/Pixabay, 6, 8, 10, 12, 14, 16; ©Jaen0000/Pixabay, 6, 8, 10, 12, 14, 16; Various images throughout courtesy of Scratch

Library of Congress Cataloging-in-Publication Data

Names: Fontichiaro, Kristin, author. | van Lent, Colleen, author.
Title: Coding in the city / by Kristin Fontichiaro and Colleen van Lent.
Description: Ann Arbor, Michigan : Cherry Lake Publishing, 2020. | Series: Operation code | Includes bibliographical references and index. | Audience: Grades 2-3.
Identifiers: LCCN 2019035733 (print) | LCCN 2019035734 (ebook) | ISBN 9781534159259 (hardcover) | ISBN 9781534161559 (paperback) | ISBN 9781534160408 (pdf) | ISBN 9781534162709 (ebook)
Subjects: LCSH: Scratch (Computer program language)–Juvenile literature. | Computer programming–Juvenile literature. | Cities and towns–Computer simulation–Juvenile literature.
Classification: LCC QA76.73.S345 F656 2020 (print) | LCC QA76.73.S345 (ebook) | DDC 005.13/3–dc23
LC record available at https://lccn.loc.gov/2019035733
LC ebook record available at https://lccn.loc.gov/2019035734

Cherry Lake Publishing would like to acknowledge the work of the Partnership for 21st Century Learning, a Network of Battelle for Kids. Please visit *http://www.battelleforkids.org/networks/p21* for more information.

Printed in the United States of America
Corporate Graphics

NOTE TO READERS: Use this book to practice your Scratch 3 coding skills. If you have never used Scratch before, ask a parent, teacher, or librarian to help you set up an account at *https://scratch.mit.edu*. Read the tutorials on the website to learn how Scratch works. Then you will be ready for the activities in this book! You will practice using variables, if/then statements, copying code to other sprites, using effects to change a sprite's look, and more! Find all the starter and final programs at *https://scratch.mit.edu/users/CherryLakeCoding*.

Table of Contents

Help Me Advertise Cherry Lake City!

I'm Mayor Brown. I need your help. Our city is great, and I want everybody to know about it. Can you help me make a **slide show** of our favorite spots?

I got things started at *https://scratch.mit.edu/projects/317548181*. Can you help me with the rest?

Pro Tip!

Scratch lets you see other people's projects and copy them to make them your own. My code contains the starter backdrops we will use in this book.

Backdrops tab

This area shows you which backdrop you are using

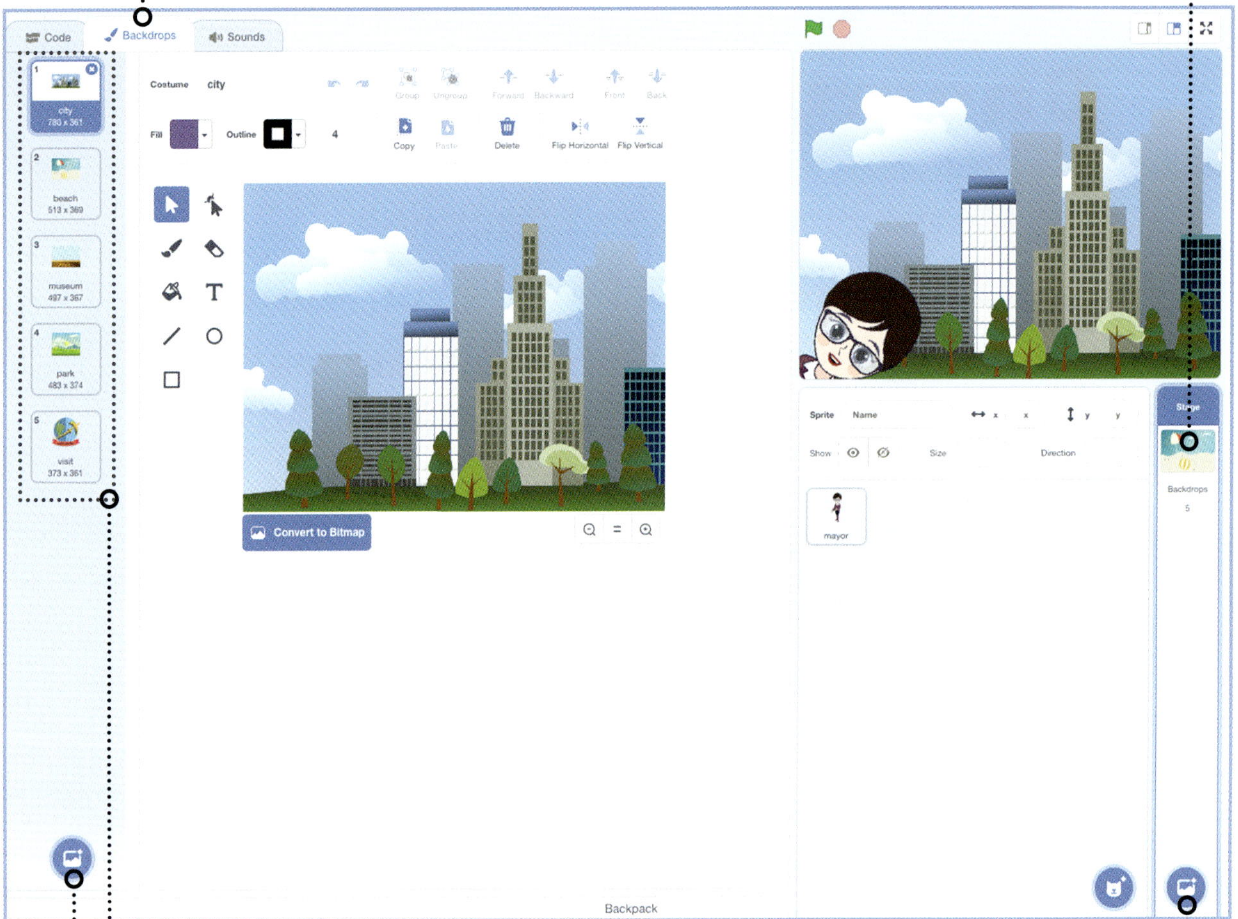

Find the preloaded backdrops here

In future projects, click one of these **icons** to add more backdrops

Getting to Know Backdrops

Did you know you can change the background of your Scratch program? You do that by changing the backdrops. We put five backdrops in the program. We need you to finish three!

To find the preloaded backdrops, click the area that shows you which backdrop you are using. Then click the Backdrops tab.

Pro Tip!

In future projects, you can use Scratch's built-in backdrops, **upload** your own, get a surprise backdrop, or make a new backdrop from scratch.

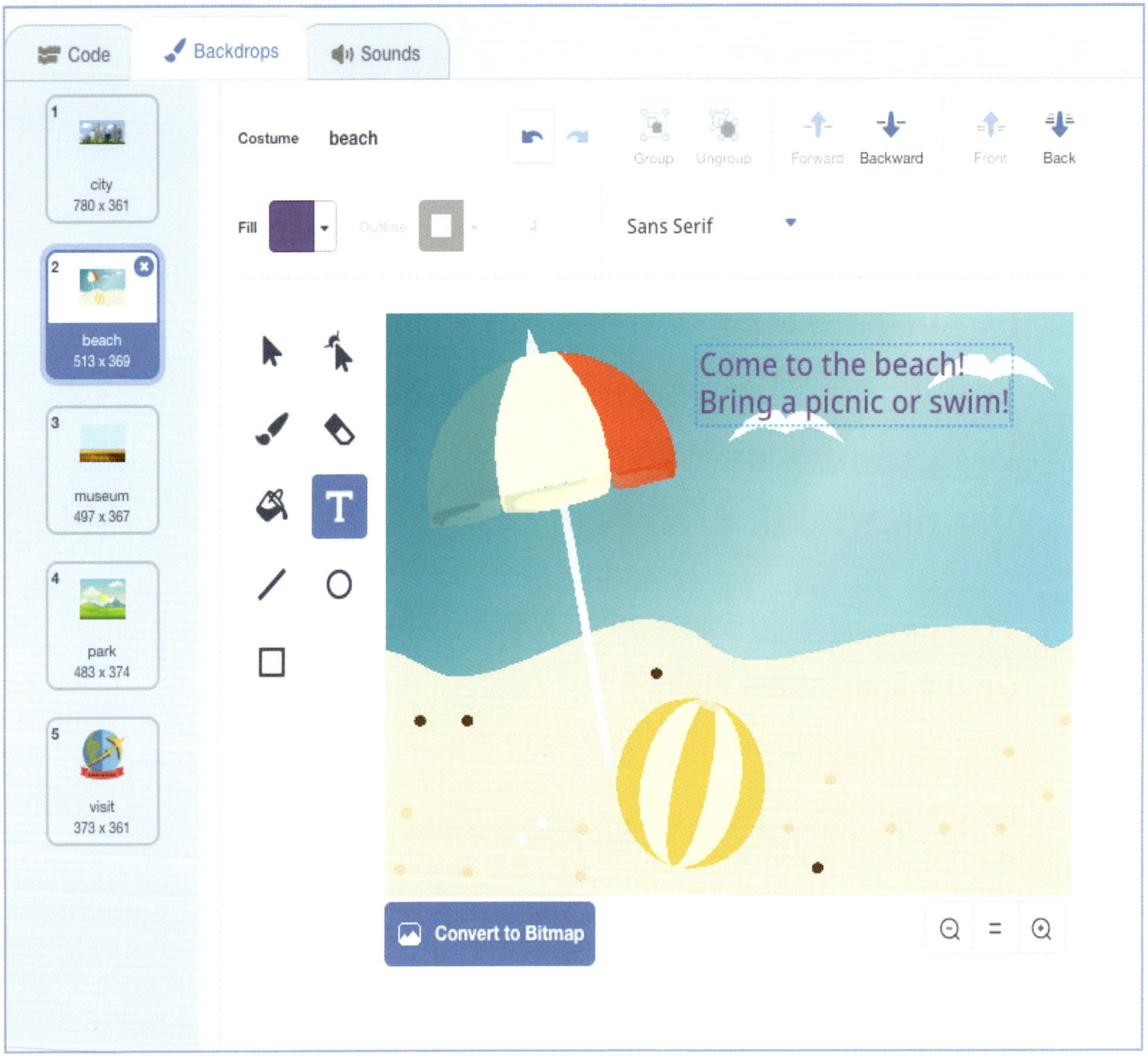

Code
Backdrops
Sounds
1
city
780 x 361
2
beach
513 x 369
3
museum
497 x 367
4
park
483 x 374
5
visit
373 x 361
Costume
beach
Group
Ungroup
Forward
Backward
Front
Back
Fill
Outline
Sans Serif
Come to the beach!
Bring a picnic or swim!
Convert to Bitmap

Adding Words to a Backdrop

You can add words to a backdrop. This is helpful when you code games and want to say, "Game over!"

Here, we want to tell people why the beach is special. Click on the beach backdrop.

Click on the  icon. Then click on the backdrop where you want your words to go. Now type!

Pro Tip!

If you see a blue box that says "Convert to **Vector**," click it so that your tool layout matches this one. Once that happens, the blue box will change to "Convert to Bitmap."

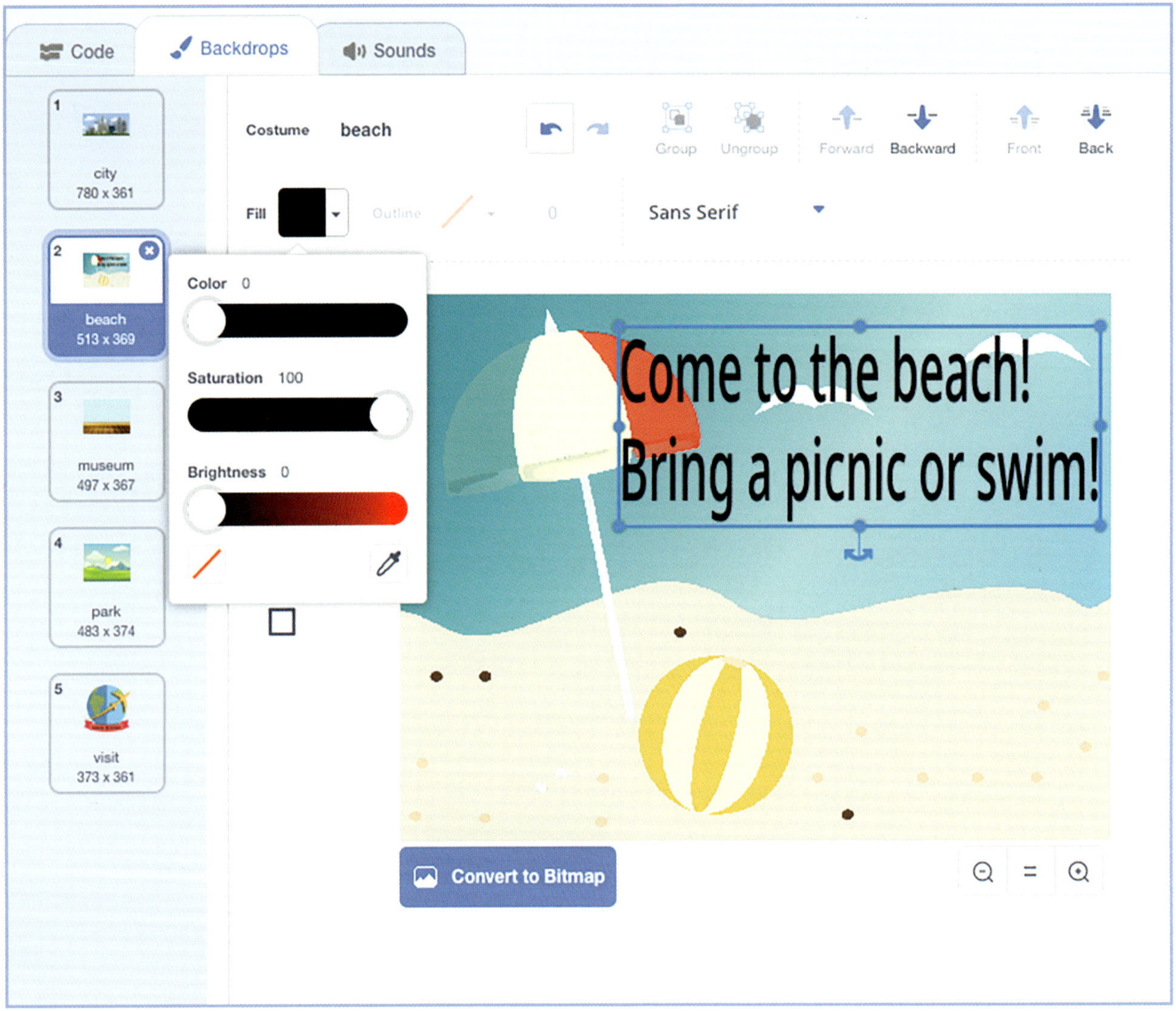
Code
Backdrops
Sounds
1
city
780 x 361
2
beach
513 x 369
3
museum
497 x 367
4
park
483 x 374
5
visit
373 x 361
Costume
beach
Group
Ungroup
Forward
Backward
Front
Back
Fill
Outline
0
Sans Serif
Color 0
Saturation 100
Brightness 0
Come to the beach!
Bring a picnic or swim!
Convert to Bitmap

Making Your Text Fancy

Let's change the color of your words.

Click on the **Fill** icon. A set of **sliders** will pop up. Move them back and forth to find the color you want. This can be a bit tricky.

Pro Tip!

To make the words bigger or smaller, drag the text box's dots up, down, or side to side.

Change your line's color here

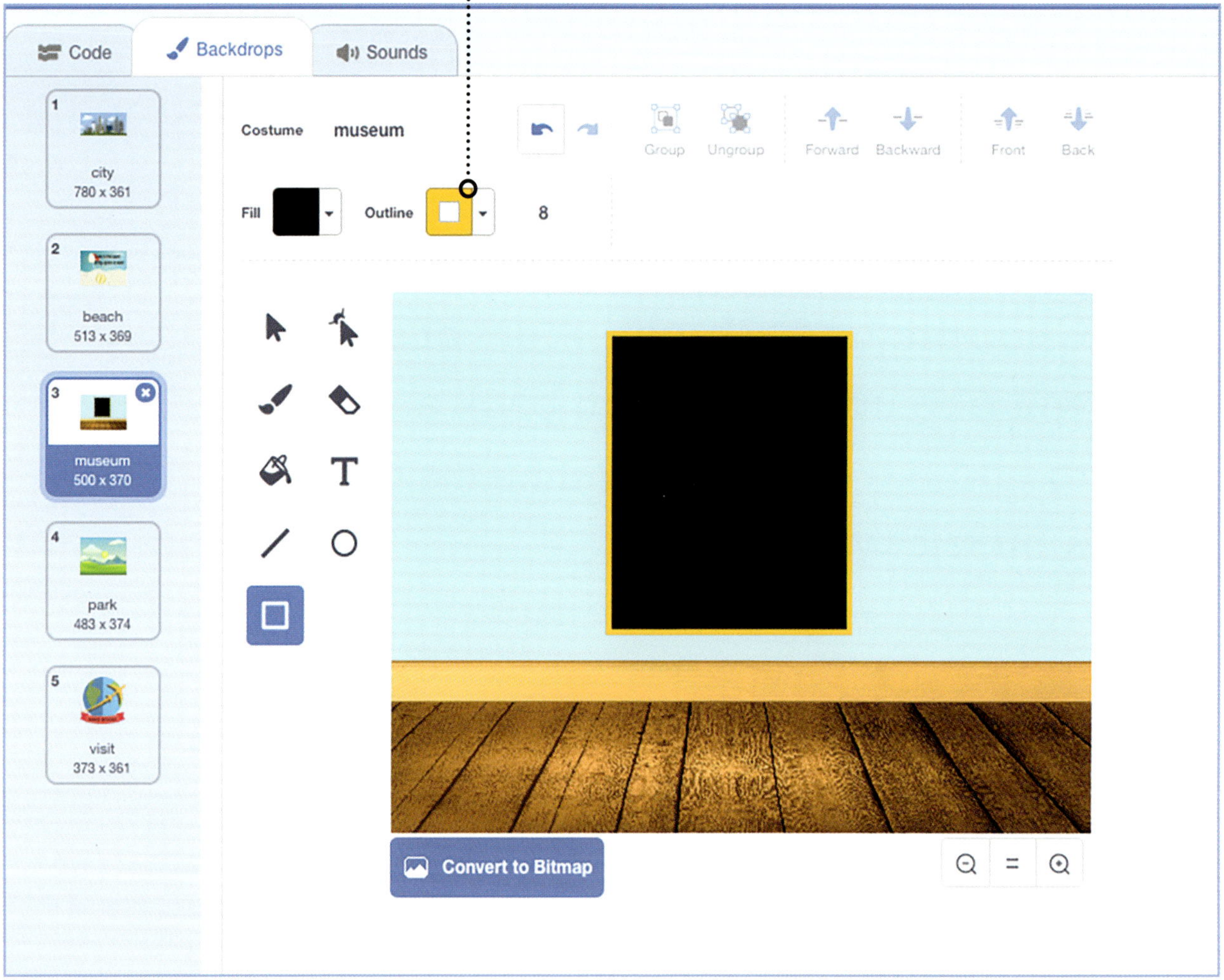

Adding Art to the Museum

We need art in the museum! Let's switch to the next backdrop.

To start, click on the ▢ icon. Then click and drag until you have a rectangular shape.

You need to make two color choices: Fill (for the painting) and **Outline** (for the frame). Click on each option and choose colors for your art.

Pro Tip!

See the number next to Outline? That number says how wide the line should be. The **default** is 1. We used 8. Make sure you click on the shape you drew before you adjust the Fill and Outline. Otherwise, Scratch won't know what to change!

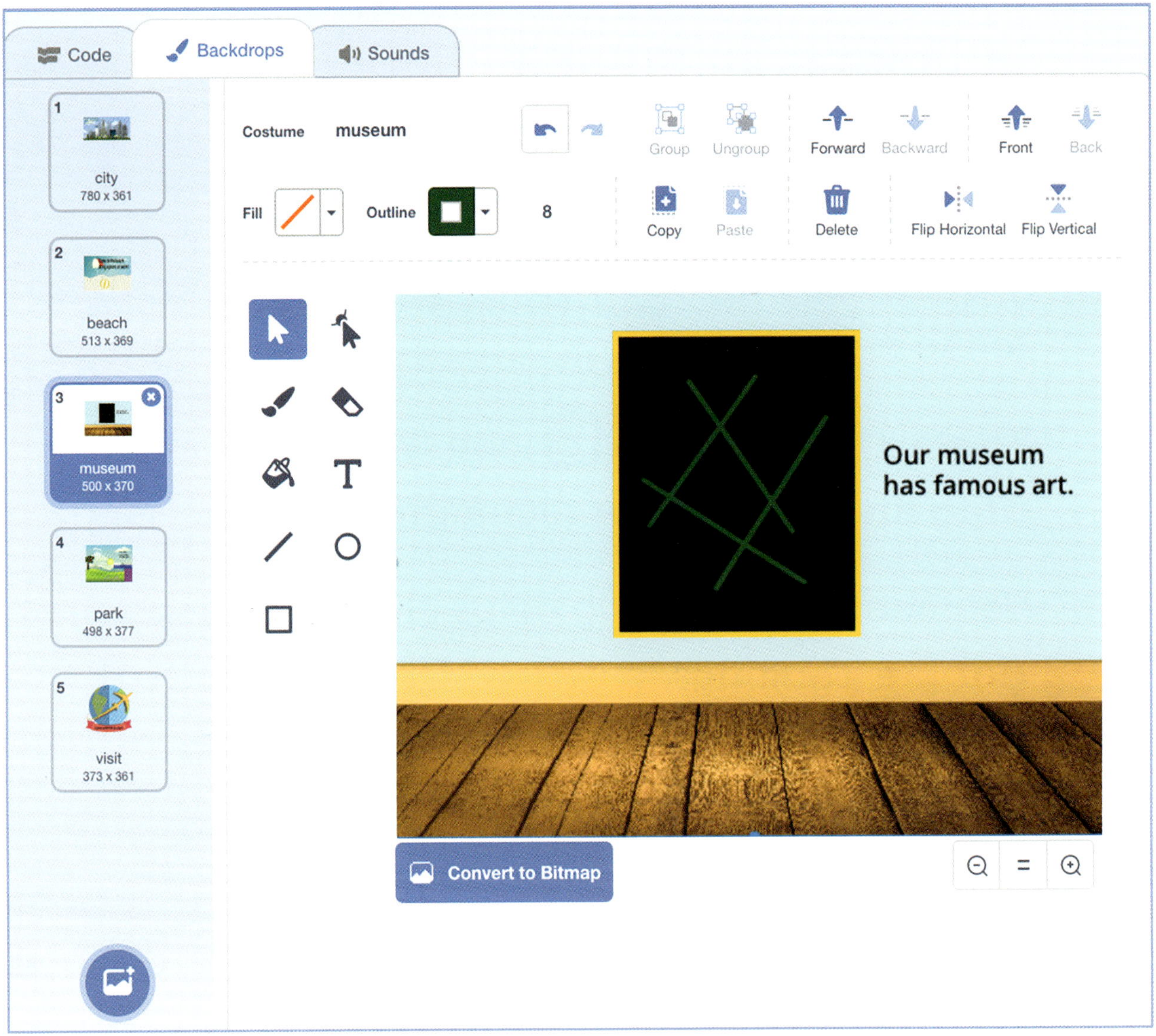
Code
Backdrops
Sounds
1
city
780 x 361
2
beach
513 x 369
3
museum
500 x 370
4
park
498 x 377
5
visit
373 x 361
Costume
museum
Group
Ungroup
Forward
Backward
Front
Back
Fill
Outline
8
Copy
Paste
Delete
Flip Horizontal
Flip Vertical
Our museum
has famous art.
Convert to Bitmap

Adding Modern Art

We need to add art to the painting.

Click on the / icon. Make sure the line is the color you want. Then click the area where you want the line to start. Drag until you have the line you want.

Add more lines to your art piece. Maybe also change the line color or thickness using the Outline option.

Add text to describe the museum.

Pro Tip!

If you decide later to **delete** a line, click on the ↖ icon first. Next, click on the line you want to delete. Then, hit the delete key on your keyboard.

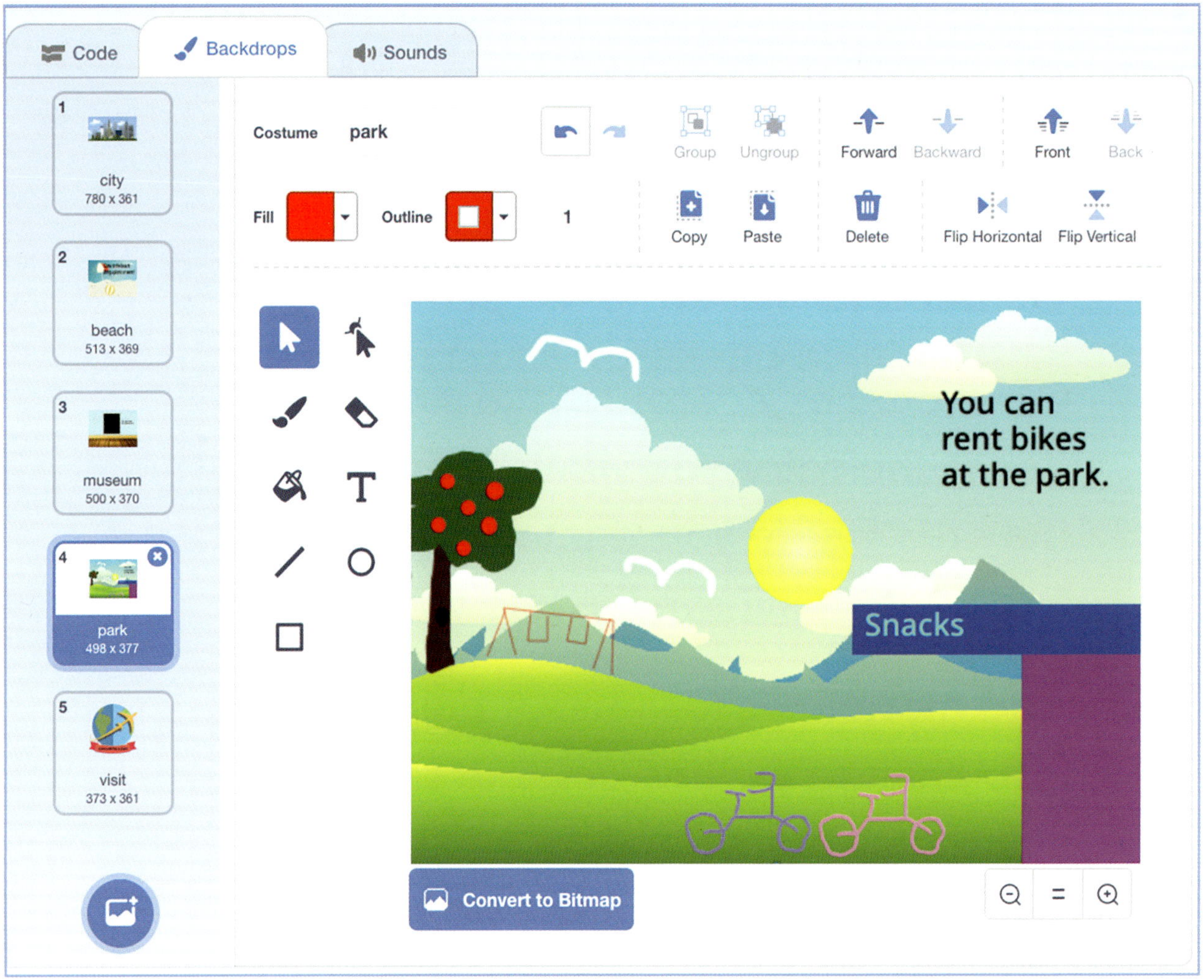
Code
Backdrops
Sounds
1
city
780 x 361
2
beach
513 x 369
3
museum
500 x 370
4
park
498 x 377
5
visit
373 x 361
Costume
park
Group
Ungroup
Forward
Backward
Front
Back
Fill
Outline
1
Copy
Paste
Delete
Flip Horizontal
Flip Vertical
You can rent bikes at the park.
Snacks
Convert to Bitmap

The Park Is Empty!

We need your help to design the park. Find the park backdrop and start drawing!

Here are a few ideas to help you get started:

- Use the icon to make a swing set.
- Use the icon to draw a building, like a snack shack.
- Use the icon to create curvy lines for birds, bikes, or trees.

Pro Tip!

Need to erase something? Try one of these:

- Click on the icon and drag the mouse over what you want to get rid of.
- Click on the icon. Then click on your object and click the delete key on your keyboard.
- Click on the icon. Then click the trash can icon.

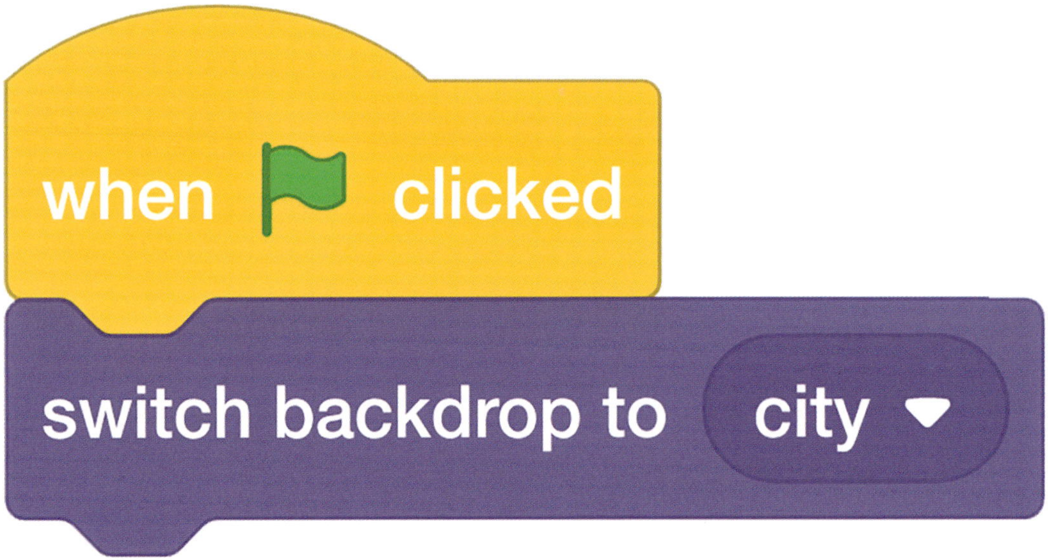
when clicked
switch backdrop to
city

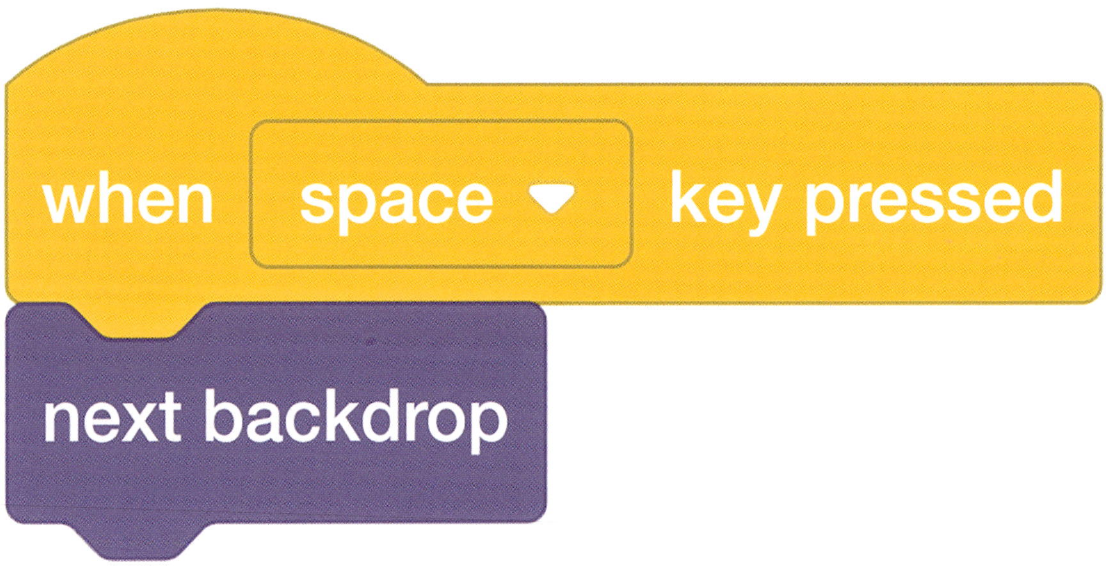
when
space
key pressed
next backdrop

Coding the Slide Show

Now that our backdrops look great, we are ready to add code to them.

Click the Code tab to get back to the main screen. Then, click on my Mayor sprite. We will add this code to my sprite.

This code makes sure that we always start with the city backdrop. Every time a visitor presses the space key, the next backdrop will show up.

Pro Tip!

Good coders always test their Scratch code at least twice before being certain that it works. Try it!

You Did It!

You helped our city look great. I know more tourists will visit now.

Have fun trying ideas like these:

- Add another location.
- Add raindrops or snowflakes to show different seasons on the backdrops.
- Change to a different tour guide.
- Add **sprites** for decoration.

What other ideas do you have?

Pro Tip!

If you would like to see our final code, go to *https://scratch.mit.edu/projects/316833914*.

Glossary

default (DEE-fawlt) the choice Scratch makes for you in advance that you can change

delete (dih-LEET) to remove something from your screen

fill (FIL) color that goes inside of a shape

icons (EYE-kahns) little pictures that you can click on

outline (OUT-line) a line that goes around the outside edge of a shape

sliders (SLIDE-urz) knobs or levers you can move left to right to change settings

slide show (SLIDE SHOH) a series of images

sprites (SPRYTS) characters or objects in Scratch

upload (UHP-lohd) to copy a file from your computer to Scratch, the Web, or another software

vector (VEK-tur) a kind of graphic that can be made bigger or smaller while still looking smooth on its edges

Find Out More

Books

LEAD Project. *Super Scratch Programming Adventure!* San Francisco, California: No Starch Press, 2019.

Lovett, Amber. *Coding with Blockly.* Ann Arbor, Michigan: Cherry Lake Publishing, 2017.

Websites

Scratch

http://scratch.mit.edu

Build your Scratch code online at this site.

Scratch Wiki

https://en.scratch-wiki.info

If you get stuck, ask an adult to help you look on this site for advice.

Index

About the Authors

Kristin Fontichiaro teaches at the University of Michigan School of Information. She likes working with kids on creative projects from coding to sewing to junk box inventions. She has written or edited almost 100 books for kids.

Colleen van Lent teaches coding and Web design at the University of Michigan School of Information. She has three cool kids and a dog named Bacon. She wishes she could touch her toes.